GLUTATAMIN

Mastering Its Role In Health, Beauty, And Vitality - Key Insights For A Balanced Life On Optimizing Wellness With Glutathione

SAMANTHA ZYLAR

Contents

CHAPTER ONE

Overview

One substance that is essential to many biological functions in the human body is glutamate. Known by several names, glutathione is an extremely potent antioxidant present in all cells, tissues, and organs.

The maintenance of optimal health and the appropriate operation of numerous body systems depend on glutathione. We shall explore the science of Glutatamin and why it is important for your health in this succinct synopsis.

The Mechanisms Of Glutatamin

Three amino acids make up the tripeptide molecule glutathione, also known as gluttatamin: glutamine, glycine, and cysteine. It works as a powerful detoxifier and antioxidant to shield cells from oxidative stress and damage from free radicals.

As the body's first line of defense against hazardous substances and preserving the integrity of cellular structures, glutathione functions as an antioxidant. Additionally, this molecule supports general health by assisting in the removal of pollutants and regulating the immune system.

Why Vitamin D Is Important

Vitamin B12 is essential for multiple reasons:

1. Antioxidant Protection: It is essential for lowering oxidative stress and safeguarding cells against harm. This is crucial for delaying the aging process and preventing a number of chronic disorders.

2. Detoxification: Glutatamin is essential for the body's detoxification because it binds to and gets rid of

dangerous poisons, heavy metals, and other elements that are damaging to health.

3. Immune Support: It strengthens the immune system's defenses against diseases and infections.

4. Cellular Health: Glutatamin helps keep cells healthy and working properly so they can carry out their tasks as best they can.

5. Chronic Illness Prevention: An increasing amount of evidence indicates that keeping glutathione levels in check may lower the chance of developing a number of chronic illnesses, such as cancer, heart disease, and neurological conditions.

In conclusion, glutathione, also known as glutamate, is an essential substance for human health. Because of its capacity to mitigate oxidative stress and aid in the detoxification processes, it is vital for preserving health and averting a variety of ailments. Knowing the science behind glutamin is essential to helping you make wise decisions regarding your diet, lifestyle, and supplementation so that your body has what it needs to stay healthy.

Fundamentals Of Glutamine

Describing Glutamine: As one of the twenty amino acids present in proteins, glutamine serves as a fundamental component of protein in the human body. Since it is a non-essential amino acid, the body is able to manufacture it on its own. Glutamate is conditionally essential because the body's need for it may outweigh its capacity to create it under specific circumstances, such as disease or extreme physical stress.

The Function Of Glutamine In The Body:

Glutamine is essential to the body in a number of ways.

1. A vital step in the process by which the body creates and repairs tissues is

protein synthesis. This is especially crucial for the development and repair of muscles.

2. Support for the Immune System: Glutamine serves as a vital source of energy for immune cells, bolstering the body's resistance to infections and illnesses.

3. Gut Health: It helps to preserve the integrity and functionality of the gut by serving as the main energy source for the cells lining the digestive tract.

4. Ammonia Detoxification: Glutamine aids in the body's removal of surplus ammonia, a waste product of protein synthesis that becomes poisonous if left to build up.

5. Brain Function: Glutamine plays a part in brain function by acting as a precursor for the neurotransmitter glutamate and by crossing the blood-brain barrier.

Sources Of Glutamine:

Glutamine can be obtained through a variety of dietary sources, such as:

1. Foods High in Protein: Glutaamine can be found in meat, fish, poultry, dairy products, and eggs.

2. Plant-Based Sources: Those following a vegetarian or vegan diet can obtain glutamine from legumes, tofu, spinach, and cabbage.

3.	Supplements:	Athletes, bodybuilders, and others with particular medical conditions frequently use glutamine supplements, which are also readily available.

4. Bone Broth: Because it has glutamine among other amino acids, some people drink bone broth.

In conclusion, glutamine is a necessary amino acid that supports the body's immune system, gastrointestinal tract, and muscle growth, among other processes. Although the body normally synthesizes it, there are circumstances in which taking supplements or consuming food sources can be helpful.

Glutamine And The Health Of Cells

An important amino acid that is vital to cellular health is glutamine, not glutamin. Here is a quick summary of its importance in terms of immune system support, oxidative stress protection for cells, and antioxidant qualities:

1. Glutamine And Cellular Health: One of the most prevalent amino acids in the human body, glutamine plays a critical role in preserving the general health of cells.

Because it is an essential component of proteins, it is necessary for a number of cellular functions, tissue healing, and growth.

2. Counteract The Negative Effects:

The antioxidant qualities of glutamine can be used to counteract the negative effects of free radicals in the body. Unstable chemicals known as free radicals have the ability to harm DNA and biological components. Glutamate can lessen this damage by functioning as an antioxidant, enhancing the general health of cells.

3. Defending Cells Against Oxidative Stress

Stress: Oxidative stress is a potentially harmful condition that arises when the body's equilibrium between antioxidants and free radicals is upset. By taking part in the body's antioxidant defense processes, glutamine can assist in reducing the effects of oxidative stress. This defense is essential for avoiding cellular deterioration and preserving the healthy operation of tissues and organs.

4. Boosting The Immune System:

Glutamine is necessary for the immune system to operate correctly. It is involved in the manufacture of several immune-related chemicals and provides immune cells with energy.

Because the body needs a strong immune system to fight off infections and illnesses, glutamine's function in immune support is essential to cellular health.

In conclusion, glutamine is an essential amino acid that supports the immune system, protects against oxidative stress, and offers antioxidant protection for cells.

 Its complex function in preserving general health emphasizes how vital it is to the body's physiological functions.

CHAPTER TWO

Vitamin D In The Detox Process

The tripeptide molecule glutathione, which is made up of the amino acids cysteine, glutamic acid, and glycine, is essential for the removal of toxins. It functions as a vital antioxidant and is crucial to the body's detoxification processes.

Heavy metals, contaminants, free radicals, and other toxic chemicals are among the things that glutathione aids in eliminating. This procedure is essential for preserving general health and averting certain illnesses.

How The Liver Uses Glutathione:

The body's detoxification process primarily involves the liver. Liver cells have particularly large amounts of glutathione. By taking part in several stages of detoxification, it helps in the breakdown and elimination of dangerous substances.

1. Phase I Detoxification: Glutathione aids in the first stage of detoxification by enabling the transformation of toxins that are soluble in fat into substances that are soluble in water. The body can eliminate the poisons more easily as a result of this change.

2. Phase II Detoxification: Glutathione is essential for this stage because it binds to hazardous chemicals and reduces their toxicity. It combines with toxins to generate conjugates, which the body excretes through the bile and urine.

Glutathione also functions as a potent antioxidant, protecting cells against oxidative stress and harm from free radicals. It increases the body's capacity to neutralize dangerous substances by aiding in the regeneration of other antioxidants like vitamins C and E.

Benefits Of Detoxification:

1. Enhanced Immune Response: A strong immune system depends on a detoxifying mechanism that is supported by glutathione. It assists the body in fighting off illnesses and infections by getting rid of things that could weaken the immune system.

2. Decreased Oxidative Stress: Glutathione contributes to the body's ability to regulate the amount of reactive oxygen species (ROS). Elevated levels of oxidative stress have been linked to a number of chronic illnesses, including cancer,

neurological conditions, and heart issues.

3. Protection Against Environmental Toxins: Glutathione's detoxifying properties are essential for reducing the harmful effects of exposure to chemicals, heavy metals, and pollutants in a period of rising environmental pollution.

4. Cellular Health: Glutathione promotes general health by protecting the integrity and health of cells. It promotes healthy cell activity and lessens the symptoms of aging.

5. Liver Function: Because the liver is the body's main organ for detoxification, glutathione is

necessary for it to function at its best. Sustaining sufficient amounts of glutathione promotes liver function and its capacity to eliminate toxins.

6. Athletic Performance: By decreasing oxidative stress and eliminating lactic acid, glutathione is used by certain fitness enthusiasts and athletes to assist in lessening muscular soreness and expediting recovery.

In conclusion, glutathione plays a vital role in the body's detoxification process. It promotes general health and well-being by assisting in the elimination of pollutants and offering defense against oxidative stress. For detoxification and general health, maintaining enough levels of glutathione through a healthy diet, exercise, and, in certain situations, supplementation, might be helpful.

Glutamine In The Prevention Of Diseases

One important amino acid that is necessary for preventing disease is glutamine. It is especially important in:

1. **Glutamine And Chronic Diseases:** By improving immune cell activity, glutamine strengthens the immune system and helps avoid chronic diseases. It also helps to support general health, lower inflammation, and preserve the integrity of the intestinal barrier. Reduced risks of long-term illnesses like diabetes, obesity, and autoimmune diseases may result from these effects.

2. The Function Of Glutamine In Heart Health:

By lowering oxidative stress and inflammation, glutamine contributes to the upkeep of a healthy cardiovascular system. By stopping the onset of atherosclerosis, a major risk factor for heart disease, this amino acid can promote heart health.

3. Glutamine And Cancer Prevention: There are several ways that glutamine can help prevent cancer. Although glutamine is an essential fuel for rapidly proliferating cells, including cancer cells, some research indicates that restricting glutamine consumption under specific conditions may prevent cancer from spreading. To completely comprehend the connection between glutamine and cancer prevention, more study is necessary.

In conclusion, glutamine is critical for preserving general health, and it may also help heart health, prevent chronic illnesses, and prevent cancer, though its exact role in preventing cancer is still up for discussion and investigation.

Vitamins And Skin Conditions

Glutamine And Hygienic Skin

An important amino acid that is necessary for many facets of skin care and upkeep is glutamine. It is the main building block of collagen, a protein that gives skin its strength, elasticity, and structure. Glutamine plays a pivotal role in multiple essential processes that lead to radiant and healthy skin.

1. Production of Collagen: One of the structural proteins that give skin suppleness and firmness is collagen, and one of its building blocks is glutamine. Enough collagen is needed to keep skin looking young and avoid wrinkles and sagging. Skin health can be enhanced by a diet high in foods or supplements that include glutamine, which can support the manufacture of collagen.

2. Skin Regeneration and Repair: Glutamine is well-known for its function in tissue repair and wound healing. It facilitates the body's

quicker healing from wounds, including burns, cuts, and scars on the skin. Because of this, glutamine plays a crucial role in skin cell regeneration and repair, giving the skin a more youthful appearance.

3. Antioxidant Properties: Glutamine functions as a precursor to glutathione, the master antioxidant that helps shield the skin from free radical-induced oxidative stress. Age spots, wrinkles, and fine lines on the skin can all be brought on by oxidative stress. Counteracting these effects and maintaining youthful-looking skin can be achieved by maintaining enough levels of

glutathione, either by glutamine ingestion or other methods.

Skin Aging And Glutamine:

Skin aging is a multifaceted process that is impacted by a number of variables, such as environment, lifestyle, and genetics. In order to mitigate the effects of skin aging, glutamine may help in the following ways:

1. Antioxidant Defense: Glutamine, as previously noted, aids in the synthesis of glutathione, a potent antioxidant. Glutathione preserves a young appearance by scavenging free

radicals, which cause premature aging.

2. Collagen Preservation: Glutamate can aid in preventing wrinkles and the loss of skin suppleness by promoting the synthesis of collagen. As we age, collagen production naturally diminishes, but keeping proper levels of glutamine in the body can help slow down this process.

Glutamine In Skincare Products:

The beauty and cosmetics business typically integrates several amino acids, including glutamine, in their formulas. These products seek to promote skin health and address specific ailments. Glutamine may be found in serums, creams, and lotions aimed to increase skin suppleness, decrease wrinkles, and promote a healthy complexion.

It's crucial to remember that individual outcomes may vary and that glutamine's efficacy in skincare products might vary as well. The performance of the product can be greatly influenced by the formulation and additional active components used with it.

Getting Radiant, Healthful Skin:

Maintaining healthy, radiant skin requires a holistic strategy that takes into account both external and internal variables. Here are a few pointers:

1. Eat a diet high in foods high in glutamine, such as lean meats, fish, eggs, dairy products, and plant-based meals like spinach and beans. Maintain a well-balanced diet.

2. Hydration: To keep your skin moisturized and retain its elasticity, drink a lot of water.

3. Skincare Routine: Establish a skincare program that includes gentle washing, moisturizing, and sun protection. Look for goods that may

include helpful amino acids, like glutamine.

4. Sun Protection: Protect your skin from the harmful effects of UV radiation by using sunscreen and wearing protective gear.

5. Lifestyle: Avoid smoking, limit alcohol consumption, and get regular exercise to promote overall health, which reflects on your skin.

6. Supplements: Individual needs may differ, so speak with a healthcare provider before taking glutamine or other supplements to address particular skin concerns.

In summary, glutamine has the potential to improve skin health and counteract the signs of aging, mainly through its antioxidant and collagen-producing properties. Incorporating glutamine-containing foods into your diet and choosing skincare products thoughtfully can help you achieve and maintain healthy, glowing skin. However, it's vital to note that general lifestyle choices, sun protection, and a full skincare routine are equally important in the goal of attractive and youthful skin.

CHAPTER THREE

Glutatamin And Performance During Exercise

Glutamine And Physical Endurance: During physical exertion, the amino acid glutamine is essential for sustaining physical endurance. It supports overall immune function by acting as the main source of energy for immune cells. This is crucial for athletes since prolonged, rigorous training can compromise immune function, leaving them more vulnerable to disease. Glutamate can support long-term physical endurance by strengthening immune function and lowering the chance of infections that could keep an athlete off the field.

Reducing Fatigue Induced By Exercise:

Glutamine also contributes to the reduction of fatigue induced by exercise. During extended physical exertion, glutamine levels in the body can drop dramatically. This depletion can lead to muscular weakening and an increase in general weariness. Supplementing with glutamine may help replenish these levels, potentially delaying the onset of weariness and allowing athletes to perform at a higher intensity for longer periods.

Maximizing Athletic Performance: Athletes often seek to maximize their performance, and glutamine can be a valuable tool in achieving this goal. By supporting endurance and reducing exercise-induced fatigue, glutamine can contribute to improved training sessions and competition results. However, it's important to note that individual responses to glutamine supplementation can vary, and its effects may be more pronounced in certain athletes or under specific conditions. Consulting with a

healthcare professional or sports nutritionist is advisable before incorporating glutamine into an athlete's regimen to ensure safe and effective usage.

Glutathione And Mental Health

An important amino acid, glutamine, has a number of positive effects on mental health, including improved brain function, reduced stress and anxiety, and improved cognitive function. Let's explore each of these ideas in more detail:

1. The Effects Of Glutamine On Brain Function:

• Neurotransmitter Precursor: Glutamine serves as a precursor to glutamate, a neurotransmitter essential for memory and learning among other cognitive processes. The brain's most prevalent excitatory neurotransmitter, glutamate, is necessary for neuronal communication.

• Glutamine-Glutamate Cycle: Another process that helps keep the balance of neurotransmitters is the glutamine-glutamate cycle, in which glutamine also takes part. This cycle can affect how mood is regulated and is necessary for the brain to function properly.

• Brain Energy Source: The brain is an extremely energy-demanding organ. When glucose levels are low, glutamine can give the brain an extra energy source. This is particularly useful when stress or high cognitive demands are present.

2. Reducing Anxiety And Stress:

• Stress Response: The body's reaction to stress can be regulated by glutamine. Because glutamine helps to maintain the integrity of the gastrointestinal tract, which can be compromised under stress, the demand for it may rise during times of stress.

• Production of GABA: Gamma-aminobutyric acid (GABA), a neurotransmitter with relaxing and anxiety-reducing properties, is produced from glutamine. Sufficient levels of GABA have the ability to counteract excitatory neurotransmitter hyperactivity, which is frequently linked to stress and anxiety.

• Assisting the Gut-Brain Axis: The gut and the brain communicate with each other in both directions through the gut-brain axis. The health of the gut has an impact on mental health, and glutamine can help with that. Both mood and cognitive function can benefit from gut health.

3. Benefits To Cognition:

• Neurotransmitter Balance: As a precursor to glutamate, glutamine contributes to maintaining the balance of excitatory neurotransmitters in the brain. This balance is crucial for optimal cognitive function, as an excess of excitatory neurotransmitters can lead to cognitive impairments.

• Energy and Mental Performance: Glutamine's role in supplying an alternate energy source for the brain can aid in enhancing mental performance during times of elevated cognitive demands, such as studying or problem-solving.

• Protecting Against Neurological Disorders: Imbalances in neurotransmitters are associated with various neurological disorders, and by maintaining proper neurotransmitter balance, glutamine may offer protection against conditions like depression, anxiety disorders, and neurodegenerative diseases.

While glutamine's role in mental well-being is noteworthy, it's essential to emphasize that a balanced diet, including a variety of amino acids and nutrients, is crucial for overall brain health. Consultation with a healthcare practitioner is essential before adopting any dietary supplements, including glutamine, to address specific mental health difficulties.

Vitamin Glutathione In Diet

Dietary Sources Of Glutamine: Glutamine is an amino acid that is conditionally necessary, meaning that while the body can normally manufacture sufficient amounts, there are times when it may be needed more due to illness, trauma, or strenuous exercise. Plant-based foods like beans, lentils, and spinach, as well as protein-rich foods like meat, fish, poultry, dairy products, and eggs, are good dietary sources of glutamine. The body gets the amino acids required to synthesize glutamine from these sources.

Incorporating Glutamine Into Your Diet: Including glutamine in your diet can be achieved through balanced meal planning. It's crucial to consume a range of protein-rich foods as part of your daily meals to maintain a suitable intake of glutamine. If you have special health issues or dietary

restrictions, you can visit a nutritionist or healthcare professional to customize your diet properly.

Cooking With Glutamine:

When cooking, it's crucial to preserve the glutamine content in meals. High heat and prolonged cooking durations might break down glutamine, limiting its efficacy. To keep the glutamine content in your meals, consider using low-heat cooking methods like steaming, sautéing, or microwaving, which are friendlier on amino acids. Also, try not to overcook protein sources, since this can assist in preserving glutamine levels in your food.

Conclusion

Glutamine, frequently referred to as the "conditionally essential" amino acid, offers a number of benefits that go beyond its traditional role in protein synthesis. Its multiple advantages have positioned it as a crucial role in improving overall health and wellness. Additionally, ongoing research implies that the future of glutamine is optimistic, as we continue to unearth its potential applications and the pathways via which it nourishes our bodies.

The Multifaceted Benefits Of Glutamine:

Glutamine has various vital roles in the body, making it a versatile amino acid with far-reaching effects. It is renowned for its capacity to support the immune system by improving the function of immune cells and minimizing the danger of infection. Glutamine also contributes to gut health by nourishing the cells of the intestinal lining and aiding in the repair of the mucosal barrier. Moreover, it helps in maintaining nitrogen balance, a critical element for muscle growth and recuperation. This amino acid is a vital energy source for rapidly dividing cells, such as those in the digestive and immune systems. It has also been offered as a potential treatment for several medical problems, including cancer and inflammatory diseases.

Taking Charge Of Your Wellness And Health:

Integrating glutamine into one's dietary regimen can be a significant tool for enhancing general health and well-being. It can help with muscle recovery and lower the risk of overtraining syndrome, which makes it especially advantageous for athletes and people undergoing intense physical training. Additionally, for those coping with gastrointestinal disorders, glutamine's capacity to maintain gut health may bring relief from illnesses including irritable bowel syndrome (IBS) and leaky gut. It strengthens the immune system, which helps shield against illnesses and supports those whose immune systems are weakened.

The Prospects For Glutamine Studies:

The field of glutamine's possible applications research is constantly developing. Scientists are researching its usage in many medical contexts, including cancer treatment, where it may help decrease chemotherapy-induced side effects. Additionally, as we develop a deeper understanding of the gut-brain link, glutamine's significance in improving mental health and cognitive performance is becoming a focus of research.

Furthermore, the possible applications of glutamine in metabolic and cardiovascular health continue to be studied. New studies may continue to transform our knowledge of how glutamine might be used to promote health.

To sum up, glutamine is an amino acid with several uses that have a wide range of effects on health and wellbeing. Its advantages include gastrointestinal health, immunological support, muscular rehabilitation, and more.

With ongoing research, we can anticipate further discoveries regarding its potential applications and mechanisms, making it an exciting subject in the realm of health and medicine. As we explore deeper into the world of glutamine, we discover new options for maximizing our health and well-being.

www.ingramcontent.com/pod-product-compliance
Lightning Source LLC
Chambersburg PA
CBHW060844260726
48661CB00002B/600